HOMEWORK BOOK

SERIES EDITOR: BRIAN SEAGER

GRADUATED ASSESSMENT

OCR GCSE MATHEMATICS

STAGES

9
10

SECOND EDITION

- Howard Baxter
- Michael Handbury
- John Jeskins
- Jean Matthews
- Mark Patmore

D1333928

Hodder Headline's policy is to use papers that are natural, renewable and recyclable products and made from wood grown in sustainable forests. The logging and manufacturing processes are expected to conform to the environmental regulations of the country of origin.

Orders: please contact Bookpoint Ltd, 130 Milton Park, Abingdon, Oxon OX14 4SB. Telephone: (44) 01235 827720. Fax: (44) 01235 400454. Lines are open 9 a.m. to 5 p.m., Monday to Saturday, with a 24-hour message-answering service. Visit our website at www.hoddereducation.co.uk.

First published in 2007 by
Hodder Murray, an imprint of Hodder Education,
a member of the Hodder Headline Group, an Hachette Livre UK company
338 Euston Road
London NW1 3BH

Personal Tutor CD-ROM © Howard Baxter, Michael Handbury, John Jeskins, Jean Matthews, Mark Patmore, Brian Seager, Eddie Wilde, 2007; with contributions from Andy Sturman; developed by Infuze Limited; cast: Nicolette Landau; recorded at Alchemy Soho

Impression number	10	9	8	7	6	5	4	3	2
Year	2012	2011	2010	2009	2008	2007			

Cover photo © Andy Sacks/Photographer's Choice/Getty Images
Typeset in 10pt Times Ten by Tech-Set Ltd. Gateshead, Tyne and Wear.
Printed in Great Britain by CPI Antony Rowe

A catalogue record for this title is available from the British Library.

ISBN: 978 0340 915 905

Stage 9 Contents

STAG

9

Introduction

This book contains exercises designed to be used with the Graduated Assessment for OCR GCSE Mathematics course. The work covers Stages 9 and 10 of the specification.

Each exercise matches an exercise in the Graduated Assessment for OCR GCSE Mathematics Stages 9 and 10 Student's Book. The exercises in the textbook are numbered through each chapter. For instance, in Stage 10 Chapter 7, Exercise 7.2 is on using the quadratic formula to solve equations. The corresponding homework exercise is Exercise 7.2H.

You will find that the homework exercises are generally shorter than those in the Student's Book but still cover the same mathematics. Some questions are intended to be completed without a calculator, just as in the Student's Book. These are shown with a non-calculator icon in the same way. Doing these questions without a calculator is vital practice for the non-calculator sections of the module test and the GCSE examination papers.

The Homework Book gives you the opportunity for further practice on the work undertaken in class. It is also a smaller book to carry home! If you have understood the topics, you should be able to tackle these exercises confidently as they are no harder than the ones you have done in class.

More practice helps to reinforce the ideas you have learned and makes them easier to remember at a later stage. If, however, you do forget, further help is at hand. As well as the textbook, there is also, with this book, a Personal Tutor CD-ROM. This contains worked examples on key topics to revise concepts you find difficult and consolidate your understanding. The exercises supported with examples on the Personal Tutor CD-ROM are marked with an icon.

You will find the answers to this Homework Book in the Higher Assessment Pack.

Checking answers

1 Find approximate answers to these calculations by rounding each number to 1 significant figure.

a) $0.654 \div 0.123$

b) $(7.46 - 3.59) \times 47$

c) 51.5^2

d) $61.7 \div 5.8$

e) 3.7×8.1

f) 23.127×28.4

g) 73.4×46.8

h) 87.4×5.9^2

Now use a calculator to see how close your approximations are to the correct answers.

2 Find approximate answers to these calculations by rounding each number to 1 significant figure.

a) $\dfrac{38.7 \times 7.24}{4.67}$

b) $\dfrac{54.6 + 73.8}{14.7 - 5.9}$

c) $\dfrac{17.8 \times 5.7}{39.2}$

d) $\dfrac{0.58 \times 73.4}{6.12}$

e) $\dfrac{0.058}{1.1 \times 0.93}$

STAG

9

f) $\dfrac{2\cdot5 \times 3\cdot9}{0\cdot93}$

g) $\dfrac{\sqrt{3\cdot45 \times 4\cdot56 \times 5\cdot67}}{2\cdot94^2}$

Now use a calculator to see how close your approximations are to the correct answers.

3 Find approximate answers to these calculations by rounding each number to 1 significant figure.

a) $\dfrac{8\cdot57 \times 10^4}{5\cdot98 \times 10^2}$

b) $(4\cdot7 \times 10^5) \times (5\cdot86 \times 10^3)$

c) $(8\cdot51 \times 10^{-3}) \times (7\cdot24 \times 10^6)$

d) $\dfrac{7\cdot9 \times 10^5}{6\cdot57 \times 10^3}$

e) $\dfrac{5\cdot8 \times 10^9}{(3\cdot68 \times 10^{-2}) \times (4\cdot1 \times 10^6)}$

f) $\dfrac{(9\cdot36 \times 10^3) \times (8\cdot42 \times 10^{-4})}{1\cdot6 \times 10^2}$

g) $\dfrac{4\cdot27 \times 10^6}{(6\cdot04 \times 10^{-3}) \times (4\cdot2 \times 10^4)}$

h) $\dfrac{(2\cdot77 \times 10^6) \times (4\cdot8 \times 10^{-4})}{(7\cdot7 \times 10^{-6}) \times (8\cdot15 \times 10^4)}$

i) $\dfrac{(6\cdot24 \times 10^3) \times (8\cdot78 \times 10^4)}{\sqrt{9\cdot1 \times 10^{10}}}$

Now use a calculator to see how close your approximations are to the correct answers.

AGE

)

Algebraic manipulation

Multiply out the brackets.

1 $(x - 4)(x - 5)$

2 $(x + 3)(x - 7)$

3 $(x - 8)(4x - 1)$

4 $(2x + 4)(3x + 2)$

5 $(3x + 5)(x - 2)$

6 $(4x + 3)(2x - 4)$

7 $(7x - 2)(2x - 7)$

8 $(3x - 5)(2x + 3)$

9 $(3x + 7)^2$

10 $(3a + 4b)(5a + 2b)$

11 $(2m - 3n)(3m + 2n)$

12 $(5p - 3q)(2p - q)$

13 $(a + 3b)(3a - 2b)$

14 $(3x - 4y)(2x + 3y)$

15 $(5a + 2b)(5a - 2b)$

EXERCISE 2.2H

Simplify where possible.

1 $2a^3 \times 3a^2$

2 $6x^2 \times 3x^4$

3 $5a^2b \times 4ab^3$

4 $4a^3b^4 - 2a^2b^3$

5 $5a^4b^5 - 3a^4b^5$

6 $\dfrac{4c^3 \times 7c^5}{2c^2 \times c^3}$

7 $\dfrac{(8y^6)^2 \times 3y}{12y^3}$

8 $\dfrac{4p^3q^2 \times 3p^4q^3}{6p^5q^4}$

9 $\dfrac{3pq \times 16p^3q^2}{12p^7q^3}$

10 $\dfrac{16p^4}{(2p^2)^3}$

11 $\dfrac{(2m^2n^3) \times (3m^3n^2)}{4m^4n^4}$

12 $4p^2q^3 \times 3p^3q^3 \div (3p^2q^2)^2$

13 $\dfrac{10p^8}{5p^{-5}}$

14 $10a^2b^3 \times 4a^3b^2 - 4a^5b^5$

15 $\dfrac{2b^2 \times 3b^3 \times 4b^4 \times 5b^5}{10b^{10}}$

AGE
9

EXERCISE 2.3H

Factorise these fully.

1 $6x + 3y$

2 $4a - 10b$

3 $5a + 7a^2$

4 $6x^2 - 4x$

5 $3a^2 - ab + ac$

6 $4xy + 2y^2 - y$

7 $2y + 8xy$

8 $9a^2 + 3ab$

9 $8ab - 4ab^2$

10 $a^3 + a^2b$

11 $2x^2 - x^2y$

12 $6x^3 - 15xy^2$

13 $5a^2b + 15ab^2$

14 $12abc + 15ab^2 - 3a^2c$

15 $8x^2yz + 4xy^2z + 12xyz$

EXERCISE 2.4H

Factorise these expressions.

1 $x^2 - 36$

2 $49 - y^2$

3 $9x^2 - 25$

4 $4y^2 - 9$

5 $1 - 64t^2$

STA
Ç

6 $x^2 - 121y^2$

7 $81a^2 - 16b^2$

8 $36a^2 - 25b^2$

9 $100 - 49y^2$

10 $2x^2 - 50y^2$

EXERCISE 2.5H

Factorise these expressions.

1 $x^2 + 8x + 12$

2 $2x^2 + 7x + 3$

3 $2x^2 - 5x + 2$

4 $6x^2 - 7x + 2$

5 $4x^2 - 8x + 3$

6 $4x^2 + 13x + 10$

7 $4x^2 + 21x + 5$

8 $3x^2 + 16x + 5$

9 $5x^2 - 16x + 3$

10 $6x^2 - 19x + 15$

EXERCISE 2.6H

Factorise these expressions.

1 $x^2 - x - 12$

2 $2x^2 - 5x - 3$

3 $2x^2 - 5x - 12$

4 $3x^2 + 8x - 3$

5 $6x^2 + x - 1$

6 $4x^2 - 4x - 3$

7 $8x^2 - 2x - 3$

8 $6x^2 - 5x - 1$

9 $6x^2 - 7x - 10$

10 $10x^2 + 13x - 3$

EXERCISE 2.7H

Simplify these expressions.

1 $\dfrac{20a^2b^3}{5b} \times \dfrac{b^4}{2a}$

2 $\dfrac{2x^2y}{6y^2} \times \dfrac{3y^2x}{2x^2}$

3 $\dfrac{7x}{x^2 - 5x}$

4 $\dfrac{2x^2 - 6x}{x^2 - x - 6}$

5 $\dfrac{x^2 + 2x - 3}{x^2 + 8x + 15}$

STA

6 $\dfrac{x^2 + 2x - 8}{x^3 - 2x^2}$

7 $\dfrac{x^2 - 7x + 12}{x^2 - 9}$

9 $\dfrac{3x^2 + 14x - 5}{x^2 + 7x + 10}$

8 $\dfrac{x^2 - 5x}{2x^2 - 11x + 5}$

10 $\dfrac{9x^2 - 4}{3x^2 - x - 2}$

EXERCISE 2.8H

Solve these equations.

1 $2x^2 - 8x + 6 = 0$

2 $3x^2 - 2x - 8 = 0$

3 $2x^2 - 13x - 7 = 0$

4 $5x^2 + 27x + 10 = 0$

5 $5x^2 - 23x + 12 = 0$

6 $2x^2 - 5x - 3 = 0$

7 $5x^2 + 13x - 6 = 0$

8 $3x^2 - 7x - 6 = 0$

9 $6x^2 + 19x - 20 = 0$

10 $2x^2 + 11x + 12 = 0$

Proportion and variation

EXERCISE 3.1H

1 Describe the variation in each of these situations.
Use the symbol $\propto$.

a) The volume of water, y litres, in a water butt and the number of times, x, a watering can may be filled from it.

b) The time it takes, t hours, to count the votes after an election and the number of people, y, who are counting.

c) The length of a journey, d km, and the time it takes, t hours, to complete it at a constant speed.

d) The number of rotations of a bicycle wheel, r, and the diameter of the wheel, d, when covering a fixed distance.

e) The number of pumps, p, used to empty a tank and the time taken, t.

2 Describe the variation shown in each of these tables of values.
Use the symbol $\propto$.

a)

x	4	12
y	1	3

b)

x	4	2
y	10	20

c)

x	6	3
y	5	10

STAG

9

d)

x	6	3
y	10	5

e)

x	16	32
y	150	75

EXERCISE 3.2H

1 Find a formula for each variation in Exercise 3.1H, question **2**. In each case sketch the graph.

2 Boyle's law states that the pressure of a gas, P Pascals, varies inversely with its volume, V m^3, at constant temperature.

a) If $P = 10$ Pascals when $V = 100$ m^3, find the formula.

b) Find P when $V = 500$ m^3.

You may find the example useful for Exercises 3.3H and 3.4H.

EXERCISE 3.3H

1 $y \propto x^2$ and $y = 13$ when $x = 2$.
Find y when $x = 5$.

2 $y \propto x^2$ and $y = 300$ when $x = 5$.
Find y when $x = 2$.

3 $y \propto x^3$ and $y = 108$ when $x = 3$.
Find y when $x = 2$.

4 $y \propto x^3$ and $y = 48$ when $x = 2$.
Find y when $x = 5$.

5 $y \propto \dfrac{1}{x^2}$ and $y = 2$ when $x = 2$.

Find y when $x = 8$.

AGE

)

6 $y \propto \dfrac{1}{x^2}$ and $y = 5$ when $x = 2$.

Find y when $x = 5$.

7 Describe the variation shown in each of these tables of values.
Use the symbol $\propto$.

a)

x	5	10
y	12.5	100

b)

x	2	3
y	9	4

c)

x	4	2
y	200	50

d)

x	10	2
y	5	125

e)

x	8	2
y	64	1

EXERCISE 3.4H

Find a formula for each of the questions in Exercise 3.3H and sketch
the graph.

1 For each of these relationships
  (i) state the type of proportion.
  (ii) find the formula.
  (iii) find the missing value in the table.

a)

x	1	2	4
y	0.25		16

b)

x	1	2	4
y	3	12	

c)

x	1	2	4
y	1		0.25

d)

x	5		20
y	10	2.5	0.625

e)

x	10	2.5	1
y	0.25		0.025

2 The gravitational force, F, between a satellite and the Earth is inversely proportional to the square of its distance, d, from the centre of the Earth.

The rule can be written as $F \propto \dfrac{1}{d^2}$.

a) The radius of the Earth is 6400 km.
  So, at 12 800 km above the Earth's surface d is three times its value on the Earth's surface.
  What is the effect on the force of gravity at this height?

b) How far above the Earth's surface is the satellite when the gravitational force is one quarter that on the Earth's surface?

3 As part of their training, astronauts have to experience the effects
of large accelerations, called 'G forces'. The training consists of
being placed in a pod at the end of a rotor arm which is spun at
speed.

The force experienced, F, is proportional to the square of the
speed, s, of the pod.

This relationship can be written as $F \propto s^2$.

When the pod is spinning with a speed of 6 m/s the force
experienced is 2 G.

How many Gs will be experienced when the speed is 18 m/s?

STA

4 Indices

EXERCISE 4.1H

1 Write each of these in index form.

  a) The fifth root of n

  b) $\sqrt[4]{n^3}$

Do not use your calculator for
questions **2** to **4**.

Work out these.
Give the answers as whole numbers or fractions.

2 a) 5^{-1} **b)** 5^0

  c) $25^{\frac{1}{2}}$ **d)** $125^{\frac{2}{3}}$

  e) $8^{\frac{4}{3}}$

3 a) $27^{\frac{5}{3}}$ **b)** $1000^{\frac{2}{3}}$

  c) $32^{\frac{4}{5}}$ **d)** 8^{-2}

  e) $4^{\frac{5}{2}}$

4 a) $3^2 \times 16^{\frac{1}{2}}$ **b)** $3^{-1} \times 5^{-2}$

  c) $2^{-2} + 6^0 + 5^{-1}$ **d)** $64^{\frac{2}{3}} \times 8^{-\frac{1}{3}}$

  e) $4^3 - 81^{\frac{1}{2}} + \left(\frac{1}{8}\right)^{-\frac{1}{3}}$

14

You may use a calculator for questions **5** to **6**.

Work out these.
Give the answers either exactly or correct to
3 significant figures.

5 a) $2 \cdot 16^4$ **b)** $5 \cdot 04^3$ **c)** $6 \cdot 54^{-2}$

    d) $3 \cdot 72^{\frac{3}{2}}$ **e)** $24^{\frac{1}{5}}$

6 a) $(7 \cdot 2 \times 10^3)^{\frac{1}{3}}$ **b)** $1 \cdot 6^3 \times 2 \cdot 1^{\frac{1}{2}}$ **c)** $1 \cdot 8^4 - 3 \cdot 2^3$

    d) $15^2 - 36^{\frac{3}{2}}$ **e)** $3^{\frac{1}{3}} - 0 \cdot 7^{-0.5}$

EXERCISE 4.2H

1 Write each of these numbers as a power of 2,
as simply as possible.

    a) 64   **b)** $16^{\frac{3}{4}}$ **c)** 0·125

    d) 1   **e)** $2^3 \times \sqrt{2}$ **f)** $2^n \times 2^{n+1}$

2 Write each of these numbers as powers of 2 and 3, as simply as
possible.

    a) 72   **b)** $\frac{9}{16}$ **c)** $\sqrt{6}$

    d) $12^{\frac{2}{3}}$ **e)** 6^n **f)** 18^{3n}

3 Write each of these numbers as a product of its prime factors.
Use indices where possible.

    a) 192   **b)** 1400   **c)** 12 375

4 Write each of these numbers as a product of powers of prime
numbers, using indices where possible.

    a) 288   **b)** $\sqrt{27}$ **c)** $\frac{25}{81}$

    d) 2048   **e)** 36^2 **f)** 50^n

5 Rearranging formulae

EXERCISE 5.1H

pt

Rearrange each of these equations to make the letter in square brackets the subject.

1 $V = 4\pi r^2$ $[r]$

2 $x^2 + y^2 = z^2$ $[y]$

3 $\sqrt{x} = tv$ $[x]$

4 $f\backslash= \sqrt{\dfrac{x}{y}}$ $[x]$

5 $V = \frac{4}{3}\pi r^3$ $[r]$

6 $mx^2 = y^2$ $[x]$

7 $x^2 - y^2 = p^2 - q^2$ $[x]$

8 $p^2 + q^2 = (x + y)(x - y)$ $[x]$

9 $\dfrac{p}{q} = ax^2$ $[x]$

10 $f = \sqrt{\dfrac{y}{x}}$ $[x]$

11 $\dfrac{a}{b} = \dfrac{y}{x^2}$ $[x]$

12 $T = 2\pi\sqrt{\dfrac{L}{g}}$ $[g]$

Arts, sectors and volumes

6

1 Find the arc length of each of these sectors.
   Give your answers to 3 significant figures.

a)

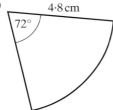

4·8 cm

72°

b)

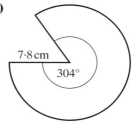

7·8 cm

304°

c)

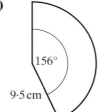

156°

9·5 cm

d)

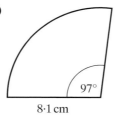

97°

8·1 cm

e)

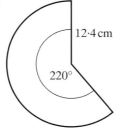

12·4 cm

220°

2 Calculate the area of each of the sectors in question **1**.
   Give your answers to 3 significant figures.

STAG
9

3 Calculate the perimeter of each of these sectors.
Give your answers to 3 significant figures.

a)

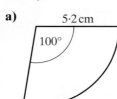

5·2 cm

100°

b)

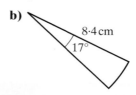

8·4 cm

17°

c)

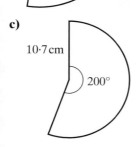

10·7 cm

200°

4 Calculate the sector angle in each of these sectors.
Give your answers to the nearest degree.

a)

5·6 cm

5·6 cm

b)

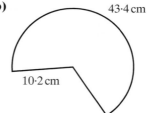

43·4 cm

10·2 cm

c)

8·4 cm

6·5 cm

d)

3·8 cm

Area = 8·2 cm²

e)

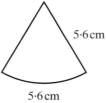

Area = 50 cm²

7·3 cm

f)

2·9 cm

Area = 15·6 cm²

5 Calculate the radius of each of these sectors.
Give your answers correct to 1 decimal place.

a)

8·4 cm

45°

b)

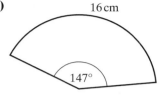

16 cm

147°

c)

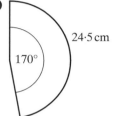

24·5 cm

170°

d)

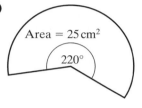

Area = 25 cm²

220°

e)

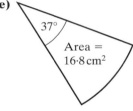

37°

Area =
16·8 cm²

f)

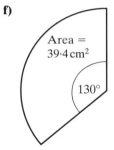

Area =
39·4 cm²

130°

6 A sector of a circle of radius 7 cm has an arc length of 18 cm.

a) Calculate the angle of the sector.

b) Calculate the area of the sector.

7 A piece of wire of length 20 cm is bent into a circular arc.

a) The angle at the centre is 30°.
What is the radius of the arc?

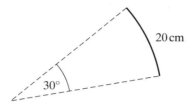

20 cm

30°

b) The same piece of wire is now bent into a circular arc with a radius of 20 cm.
What is the angle at the centre?

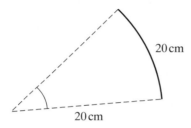

20 cm

20 cm

EXERCISE 6.2H

1 Calculate the volume of each of these pyramids.
Their bases are squares or rectangles.

a)

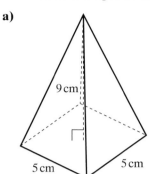

9 cm

5 cm   5 cm

b)

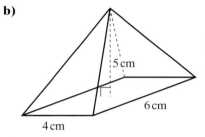

5 cm

6 cm

4 cm

c)

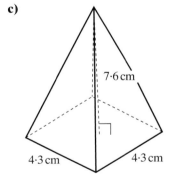

7·6 cm

4·3 cm   4·3 cm

d)

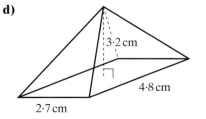

3·2 cm

4·8 cm

2·7 cm

2 Calculate the volume of each of these cones.

a)

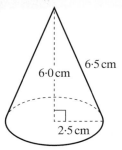

6·5 cm

6·0 cm

2·5 cm

b)

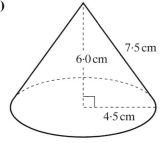

7·5 cm

6·0 cm

4·5 cm

c)

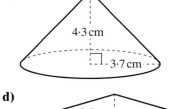

4·3 cm

3·7 cm

d)

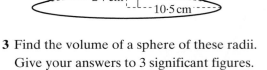

2·7 cm

10·5 cm

3 Find the volume of a sphere of these radii.
Give your answers to 3 significant figures.

a) 4.6 cm   **b)** 8 mm

4 This cylindrical jug holds 1 litre.
Its base radius is 4 cm.

Find its height.

5 This conical glass holds 150 ml.
Its top radius is 4 cm.

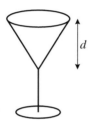

Find its depth, d.

6 A glass vase is in the shape of a cylinder with a base.
Its external height is 25 cm and its external diameter is 10 cm.
The thickness of the glass is 5 mm.
Find the volume of glass in the vase.

7 A metal sphere of radius 4 cm is melted down and then cast as a cube.
What are the dimensions of the cube?

Upper and lower bounds

1 Find the upper bound of the sum of each of these pairs of measurements.

   a) 43·2 cm and 81·7 cm
      (both to the nearest millimetre)

   b) 10·31 seconds and 19·17 seconds
      (both to the nearest hundredth of a second)

2 Find the lower bound of the sum of each pair of measurements in question **1**.

3 Find the upper bound of the difference between each of these pairs of measurements.

   a) 489 m and 526 m (both to the nearest metre)

   b) 0·728 kg and 1·026 kg (both to the nearest gram)

4 Find the lower bound of the difference between each pair of measurements in question **3**.

5 Find the upper and lower bounds of the sum of the times 3 hours 23 minutes and 1 hour 37 minutes, each correct to the nearest minute.

6 A length of timber measures 2400 mm, correct to the nearest millimetre.
Paul cuts off a piece from the end.
He intends it to be 900 mm long but his cut is only accurate to the nearest centimetre.
Give the upper and lower bounds of the length of the piece that is left.

7 One day the high tide was 2.3 m above a mark on the harbour wall.
The low tide on the same day was 2.0 m below the mark.
Both measurements were correct to 1 decimal place.
Calculate, in metres, the lower bound for the difference between
the high and low tides.

8 Two parcels weigh 247 g and 252 g, each to the nearest gram.
The cost of posting parcels increases for those weighing 500 g or
more.
If these two parcels are fastened together, will they certainly
weigh less than 500 g?
Explain your answer.

EXERCISE 7.2H

1 Find the upper bound for the area of the floor of rectangular
rooms with these dimensions.

a) 4·3 m by 6·2 m (both to 2 significant figures)

b) 4·27 m by 6·24 m (both to the nearest centimetre)

2 Find the lower bound for the area of the floor of each room in
question **1**.

3 Calculate the upper bound of the average speed in each of these
situations.
Give your answers to 4 significant figures.

a) 157 km (to the nearest kilometre) in 2·5 hours (to the nearest
0·1 hour)

b) 800·0 cm (to the nearest mm) in 103·47 seconds (to the nearest
hundredth of a second)

4 Find the lower bound of the average speed in each situation in
question **3**.

5 Calculate the lower and upper bounds of the widths of these
rectangles.

a) Area 400 cm^2 (to the nearest cm^2),
length 15 cm (to the nearest cm)

b) Area 24·5 cm^2 (to 3 significant figures),
length 5·7 cm (to 2 significant figures)

6 Calculate the upper and lower bounds of the heights of these cuboids.

    a) Volume 400 cm³ (to the nearest cm³), length 10 cm, width 5 cm (both to the nearest cm)

    b) Volume 50·0 m³ (to 3 significant figures), length 5·0 m, width 2·8 m (both to 2 significant figures)

7 Harry has eight rods, each of length 10 cm, correct to the nearest centimetre.
He places them in the shape of a rectangle as in the diagram.

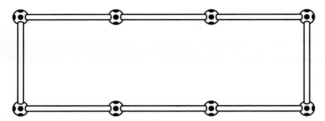

    a) What is the minimum length of the rectangle?

    b) Calculate the minimum area of the rectangle.

8 The length of Edward's training run is 380 m, correct to the nearest 10 m.
Edward completes his run at an average speed of 3·9 m/s, correct to 1 decimal place.
Calculate the upper bound of the time Edward takes to complete his run.

Similarity and enlargement

8

EXERCISE 8.1H pt

1 State the area scale factor for each of these length scale factors.

 a) 8   b) 15

2 State the volume scale factor for each of these length scale factors.

 a) 6   b) 20

3 State the length scale factor for each of these.

 a) An area scale factor of 36

 b) A volume scale factor of 8

4 An oval mirror has an area of 162 cm². What is the area of a similar mirror one and a half times as long?

5 The three tables in a nesting set of tables are similar, with heights in the ratio $1 : 1\cdot2 : 1\cdot5$. The area of the smallest table top is 120 cm². What is the area of the middle-sized table top?

6 A model of a theatre set is made to a scale of $1 : 20$. A cupboard on the model has a volume of 50 cm³. Find the volume of the cupboard on the actual set, giving your answer in m³.

7 Two jugs are similar with capacities of 1 litre and 2 litres respectively. The height of the larger jug is 14·8 cm. What is the height of the smaller jug, to the nearest millimetre?

STAG
9

8 Julia is making a model of a wooden statue.
She uses the same kind of wood as the original.
The model is on a scale of 1 : 20.

a) The statue is 15 m high.
How high will Julia's model be?

b) 500 litres of varnish were needed for the statue.
How many litres will be needed for the model?

c) Julia's model weighs 3 kg.
Estimate the weight of the statue.

9 Lego and Duplo are types of building bricks for children.
Duplo bricks are designed for younger children and are twice as
large as Lego bricks.

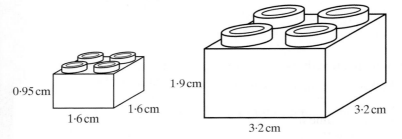

0·95 cm   1·6 cm   1·6 cm   1·9 cm   3·2 cm   3·2 cm

a) Janet makes a shape using Duplo bricks.
Her shape is 9·5 cm high.
Elaine makes a similar shape from Lego.
How high is Elaine's shape?

b) The area of the base of Elaine's shape is 64 cm².
What is the area of the base of Janet's shape?

c) The volume of Elaine's shape is 304 cm³.
What is the volume of Janet's shape?

10 This is a set of Russian dolls.

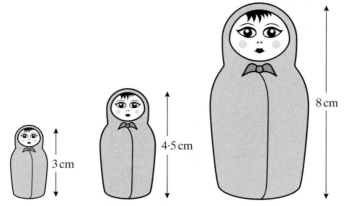

The larger dolls are enlargements of the smallest doll.
The heights of the dolls are 3 cm, 4·5 cm and 8 cm respectively.

a) The width of the smallest doll is 1·2 cm.
What is the width of the largest doll?

b) The surface area of the middle-sized doll is 8·4 cm².
What is the surface area of the smallest doll?

c) The volume of the smallest doll is 4·2 cm³.
What is the volume of the middle-sized doll?

EXERCISE 8.2H

1 Triangle ABC is mapped on to triangle A'B'C'.

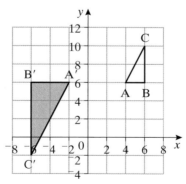

Copy the diagram and find

a) the centre of enlargement.

b) the scale factor.

2 Flag ABCD is mapped on to flag A'B'C'D'.
Copy the diagram and find

a) the centre of enlargement.

b) the scale factor.

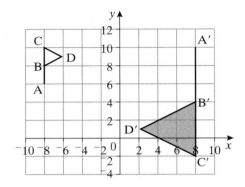

3 Draw a set of axes with the *x*-axis from 0 to 18 and the *y*-axis from 0 to 10.
Plot the points A(4, 4), B(16, 4), C(16, 8) and D(4, 8) and join them to form a rectangle.
Enlarge rectangle ABCD by a scale factor of $\frac{1}{2}$ using the origin as the centre of enlargement.

4 Draw a set of axes with both the *x*- and *y*-axes from $^-10$ to 6.
Plot the points A(2, 2), B(4, 2) and C(2, 4) and join them to form a triangle.
Enlarge triangle ABC by a scale factor of $^-3$ using the point (1, 1) as the centre of enlargement.

Probability

 EXERCISE 9.1H

1 Katie has three black pens, five blue pens and two red pens in her bag.
   She selects a pen at random.
   What is the probability that the pen is red or blue?

2 A bag contains 20 balls.
   There are six red ones and five yellow ones.

   a) A ball is selected at random.
   What is the probability that it is not red or yellow?

   b) A ball is selected at random.
   It is then replaced and another ball is selected.
   What is the probability that a red ball and then a yellow ball are selected?

3 The numbers on the menu show the probabilities that Richard chooses these dishes.

> ### *MENU*
> *First course*
> Soup   (0·6)
> Melon   (0·4)
>
> *Second course*
> Spaghetti Bolognese   (0·1)
> Lamb Biryani   (0·7)
> Chicken & Mushroom Pie (0·2)

What is the probability that Richard chooses soup and lamb biryani?

4 An ordinary dice is thrown and a coin is tossed.
   What is the probability of getting a 6 and a tail?

STAG

9

5 On any day, the probability that Jo cycles to school is 0·8.

   a) What is the probability that she does not cycle on Monday or Tuesday?

   b) Find the probability that she cycles on just one of these days.

6 From experience, the probability of Sam winning a game of Solitaire is 0·7.
   Sam plays two games.

   a) What is the probability that Sam wins the first game but loses the second?

   b) What is the probability that he loses both games?

7 On her way to work, Rami must go through a set of traffic lights and over a level crossing.
   The probability that she has to stop at the traffic lights is 0·4.
   The probability that she has to stop at the level crossing is 0·3.
   These probabilities are independent.

   a) What is the probability that Rami does not have to stop at the traffic lights or the level crossing?

   b) Rami thinks that the probability she has to stop just once is 0·28. Show why she is wrong.

8 On the way home I pass through three sets of traffic lights.
   The probability that the first set is on green is 0·5.
   The probability that the second set is on green is 0·6.
   The probability that the third set is on green is 0·75.

   Calculate the probability that

   a) I do not have to stop at any of the lights.

   b) I have to stop at at least one set of traffic lights.

   c) I have to stop at exactly two of the sets of lights.

EXERCISE 9.2H

1 A class has 8 boys and 12 girls.
Two students are selected at random from the class.

a) Draw a tree diagram to represent the two choices.

b) Calculate the probability that
    (i) they are both girls.
    (ii) one is a boy and one is a girl.

2 A hand of 13 cards contains six red cards and seven black cards.
Two cards are taken from the hand at random.

a) Draw a tree diagram to represent the two choices.

b) Calculate the probability that
    (i) both cards are black.
    (ii) at least one card is red.

3 The probability that the plum tree in my garden will produce
more than 50 kg of plums in any year is 0·6.
If the plum tree produces more than 50 kg of plums one year, then
the probability that it will produce more than 50 kg of plums in
the following year is 0·8; if it does not then the probability is 0·4.

a) Draw a tree diagram to represent the probabilities of whether
or not the tree will produce more than 50 kg of plums in two
consecutive years.

b) Calculate the probability that the tree will
    (i) produce more than 50 kg of plums in both years.
    (ii) produce more than 50 kg of plums in just one of the two
        years.

4 The probability that it will rain on Saturday is 0·3.
If it rains on Saturday then the probability that it will rain on
Sunday is 0·5.
If it doesn't rain on Saturday then the probability that it will rain
on Sunday is 0·3.

a) Draw a tree diagram to show the probability of rain during the
weekend.

b) Find the probability that it will rain on at least one day at the
weekend.

5 A box contains four red, five blue and six green counters.
Two counters are selected at random, without replacement.

Calculate the probability that

a) they are both the same colour.

b) only one of the counters is blue.

6 On his way to work, Brian has to drive through two sets of traffic lights.
The probability that the first set is green when he gets there is 0·6.
If the first set is green, then the probability that the second is also green when he gets there is 0·9, otherwise the probability is 0·2.

Find the probability that Brian will have to stop at just one set of lights.

Working in two and three dimensions

EXERCISE 10.1H

1 Find the length of each of the lines in the diagram.
  Where the answer is not exact, give your answer
  correct to 2 decimal places.

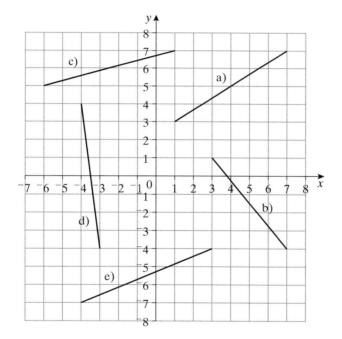

2 Find the length of the line joining each of these pairs
  of points.
  Where the answer is not exact, give your answer
  correct to 2 decimal places.

  a) A(2, 2) and B(4, 7)   **b)** C(2, 9) and D(7, 2)

  c) E(⁻2, ⁻4) and F(⁻3, 5) **d)** G(6, 3) and H(3, ⁻1)

  e) I(0, ⁻6) and J(⁻5, 6)   **f)** K(5, ⁻7) and L(⁻3, 10)

STAG

9

35

EXERCISE 10.2H

1 ABCDEFGH is a cuboid with dimensions as shown.

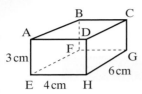

a) Calculate these.
   (i) AC
   (ii) Angle BAC
   (iii) AG
   (iv) Angle CAG

b) Take E as the origin, EH as the *x*-axis, EF as the *y*-axis and EA as the *z*-axis.
   Write down the coordinates of the vertices.

2 PABCD is a square-based pyramid with P vertically above the midpoint of the square base.
AB = 12 cm, AP = BP = CP = DP = 15 cm.
E is the midpoint of BC.
F is the midpoint of AD.

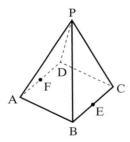

Calculate these.

a) DB

b) Angle PBD

c) PE

d) Angle PEF

3 ABCDEF is a triangular prism with angle BCF = 90°.

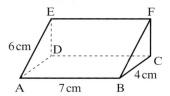

Calculate these.

a) FC

b) Angle FBC

c) EB

d) Angle EBD

4 PABCD is a square-based pyramid with P vertically above the midpoint of the square base.
AP = BP = CP = DP = 9·5 cm and AC = BD = 8·4 cm.

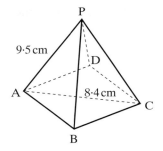

Calculate angle PAC.

5 A pencil case is a cuboid with a base measuring 6 cm by 15 cm. A pencil 17 cm long just fits in the box.

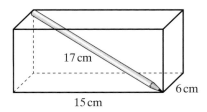

Calculate the height of the pencil case.

Working in two and three dimensions

6 The points A, B and C are in the same horizontal plane.
The angle of elevation of a vertical mast MC from A is 24·7°.
AC is 34 m and BC is 57 m.

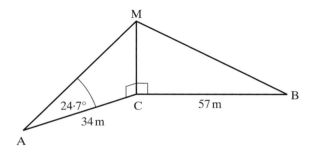

Find the angle of elevation of M from B.

EXERCISE 10.3H

1 Calculate the distance between each of these pairs of points.
Where the answer is not exact, give your answer to 2 decimal
places.

a) (1, 2, 3) and (4, 5, 6)

b) (5, 0, ⁻3) and (⁻2, 7, ⁻5)

c) (⁻1, ⁻1, ⁻1) and (1, 1, 1)

d) (⁻3, 3, ⁻3) and (4, 6, ⁻5)

e) (7, ⁻1, 4) and (⁻2, ⁻5, ⁻1)

2 The distance between a point P and a point Q with coordinates
(1, 3, ⁻2) is $\sqrt{26}$ units.
The x- and z-coordinates of P are ⁻3 and 1, respectively.
Calculate the possible y-coordinates.

EXERCISE 10.4H

1 The diagram shows the cuboid ABCDEFGH.

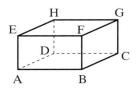

For this cuboid, sketch the triangle and label the angle between these lines and planes.

a) AG and ABCD

b) EC and BCGF

2 Given that, for the cuboid in question **1**, AB = 12 cm, BC = 8 cm and BF = 9 cm, calculate the angle between these lines and planes.

a) AG and ABCD

b) EC and BCGF

3 A cube has sides of length 10 cm.
Calculate the angle between a diagonal of the cube and a face of the cube.
Explain why your answer is true for all cubes.

4 A cuboid has a base with sides of 5·6 cm by 8·2 cm.
Its height is 4·3 cm.

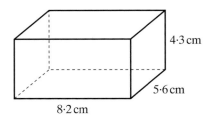

Find the angle between a diagonal of the cuboid and

a) the base.

b) a 5·6 cm by 4·3 cm face.

5 A pyramid is 8 cm high and has a square base with sides of 6 cm.
Its sloping edges are all of equal length.
Calculate the angle between a sloping edge and the base.

6 A square-based pyramid has a height of 8·3 cm.
Its sloping faces each make an angle of 55° with the base.

a) Find the length of the sloping edges of the pyramid.

b) Find the length of the sides of the base.

7 ABCDEF is a triangular wedge.
The base ABFE is a horizontal rectangle.
C is vertically above B.

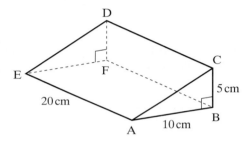

a) Calculate the length AD.

b) Calculate the angle which AD makes with
   (i)   the base ABEF.
   (ii) the face ABC.

Histograms

11

1 The table summarises the distribution of the heights of 120 children.

Height (h cm)	Number of children
$75 < h \leqslant 100$	15
$100 < h \leqslant 120$	20
$120 < h \leqslant 140$	32
$140 < h \leqslant 160$	44
$160 < h \leqslant 180$	9

Draw a histogram to show this information.

2 The table shows the numbers of men and of women in each age group at a golf club.

Age (years)	Men	Women
15–19	4	2
20–29	10	8
30–39	18	12
40–49	12	14
50–64	32	25
65–79	45	12

Show the information for men and for women on two histograms.

STAG
9

41

3 The table summarises the distribution of the money raised for charity by runners in a sponsored race.

Amount raised (£x)	Frequency
$0 < x \leqslant 50$	6
$50 < x \leqslant 100$	22
$100 < x \leqslant 200$	31
$200 < x \leqslant 500$	42
$500 < x \leqslant 1000$	15

Draw a histogram to show this information.

4 The table summarises how much the workers at a factory earn each week.

Amount earned (£x)	Frequency
$0 < w \leqslant 250$	18
$250 < w \leqslant 400$	22
$400 < w \leqslant 450$	56
$450 < w \leqslant 500$	12
$500 < w \leqslant 1000$	5

Draw a histogram to show this information.

1 This histogram shows the distribution of time spent watching TV in a week by a group of people.

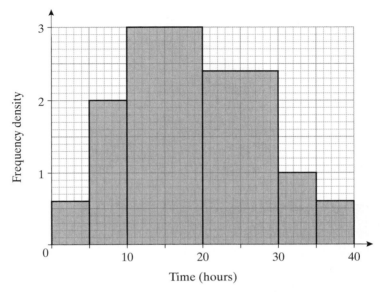

Time (hours)

There were three people who watched TV for 0 to 5 hours.

a) How many were there in each group?

b) Work out an estimate of the mean length of time spent watching TV.

2 This histogram shows the age distribution of the members of Carterknowle Methodist Church.

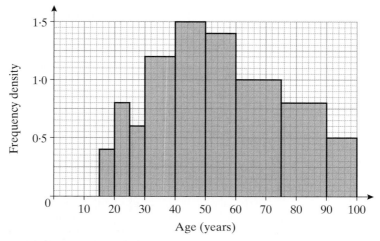

There are two members under 20.

a) How many members are there altogether?

b) Work out an estimate of the mean age of the members.

3 The histogram shows the distribution of money raised in a sponsored race.

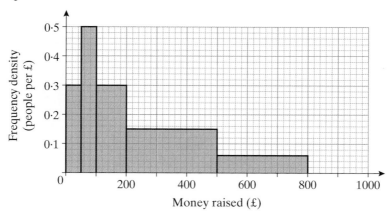

15 people raised £50 or less.

a) Calculate the frequency for each group.

b) Calculate an estimate of the total amount raised.

c) Make two comparisons with the distribution for the sponsored race in Exercise 11.1H question **3**.

4 This histogram represents a distribution of waiting times in an outpatients department one day.

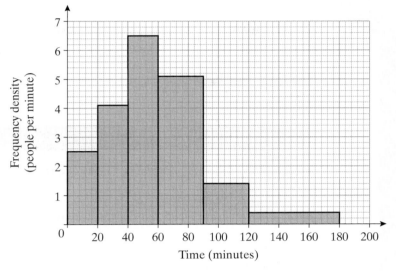

a) Make a frequency table for this distribution.

b) Calculate an estimate of the mean waiting time.

STA

12 Circle properties

You may find the example useful for
Exercises 12.1H and 12.3H.

▌▍▏ EXERCISE 12.1H

In each of the following diagrams, O is the centre of the
circle.
Find the size of each of the lettered angles.
Write down each step with the reasons for your
deductions.

1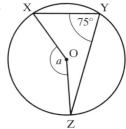

2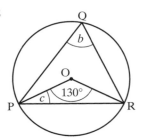

3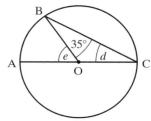

4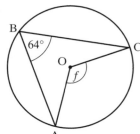

5

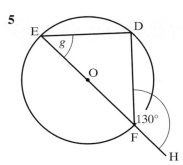

7

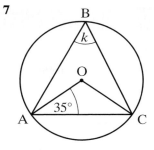

6

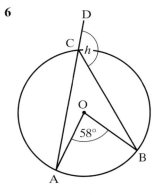

8

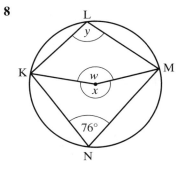

EXERCISE 12.2H

In the following questions, O is the centre of the circle.
Calculate the angles marked with letters, giving reasons for your answers.

1

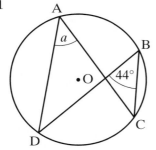

2

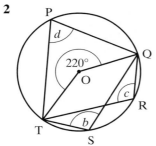

3

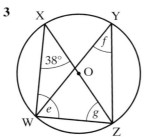

4

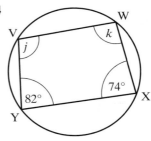

5

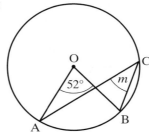

6

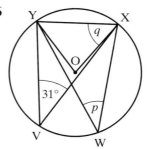

EXERCISE 12.3H

Calculate the size of each of the angles marked with a letter.
O is the centre of each circle.
Give the reasons for each step of your working.

1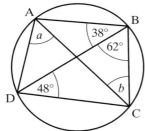

4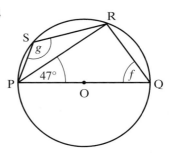

2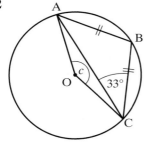

5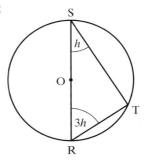

3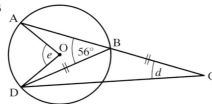

6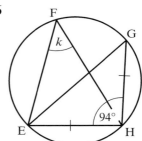

STA

EXERCISE 12.4H

In questions **1** to **4**, calculate the size of the angles marked with letters.
O is the centre of each circle.
X and Y are the points of contact of the tangents to each circle.

1

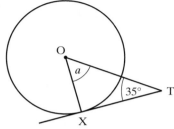

3

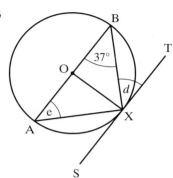

2

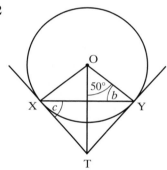

4

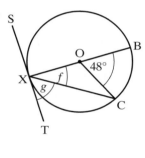

In questions **5** and **6**, find the lengths marked with letters.

5

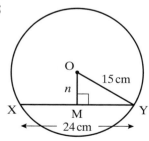

6

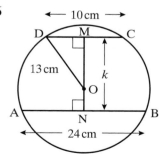

EXERCISE 12.5H

In these questions, ST is a tangent to the circle, which has centre O.
Find the lettered angles, giving reasons for your answers.

1

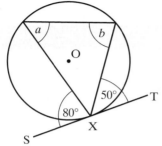

4

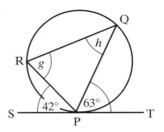

2

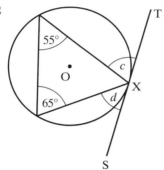

5

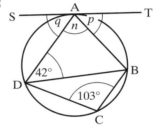

3

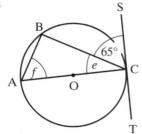

6

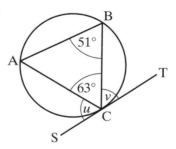

STA

Straight-line graphs

EXERCISE 13.1H

Find the equation of each of these straight lines.

1

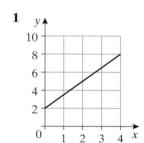

2

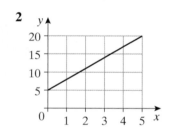

3

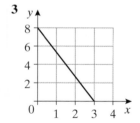

4

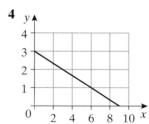

5 A line with gradient $\frac{2}{3}$ passing through the point $(2, 3)$.

6 A line with gradient $-\frac{3}{4}$ passing through the point $(3, 0)$.

7 A line passing through $(1, 4)$ and $(4, 7)$.

8 A line passing through $(2, 3)$ and $(5, 9)$.

9 A line passing through $(^-1, 5)$ and $(3, ^-7)$.

10 The table shows how the velocity of an object varies with time.

Time (t seconds)	5	10	15
Velocity (v m/s)	80	140	200

    a) Draw the graph of v against t.

    b) Find the gradient and the intercept on the v-axis.
What do they represent?

    c) Find the equation connecting t and v.

 EXERCISE 13.2H

1 Find the gradient of a line perpendicular to the line joining each of these pairs of points.

    a) $(1, 1)$ and $(5, 3)$

    b) $(1, 2)$ and $(4, ^-2)$

    c) $(^-1, 5)$ and $(2, 8)$

2 Find the equation of the line that passes through $(1, 0)$ and is parallel to $y = 2x + 6$.

3 Find the equation of the line that passes through $(2, 3)$ and is parallel to $4x + 2y = 7$.

4 a) State the gradient of the line $3x + 5y = 6$.

    b) Find the equation of the line that passes through $(1, 1)$ and is perpendicular to $3x + 5y = 6$.

5 Find the equation of the line that passes through $(4, 1)$ and is perpendicular to $y = 4x + 3$.

6 Find the equation of the line that passes through $(0, 5)$ and is perpendicular to $3y = x - 1$.

7 Find the equation of the line that passes through $(2, 5)$ and is perpendicular to $7y + 2x = 9$.

8 Which of these lines are

   a) parallel? **b)** perpendicular?

   $$y = x + 5 \quad y = 3x + 5 \quad x + 3y = 5 \quad 4x - y = 5$$

9 Two lines cross at right angles at the point $(2, 5)$.
One passes through $(4, 7)$.
What is the equation of the other line?

10 In the diagram AC is a diagonal of the square ABCD.

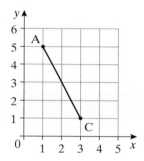

Work out

   a) the equation of the line AC.

   b) the equation of the line BD.

   c) the coordinates of B and D.

Surveys and sampling

EXERCISE 14.1H

1 You wish to investigate how regularly students at your school visit the cinema.
You decide to write a questionnaire to find out.
Write down two or more useful questions which would help you gather the necessary information.

2 A local radio station is about to start broadcasting. They decide to send out a questionnaire to find out the sort of programmes that are required.
Comment on each of these questions and where necessary write a more useful one.

a) What type of programme do you like most?

b) Do you listen to the radio in the afternoon?

c) Do you like competitions and phone-ins?

d) How much do you earn?

Are there any other questions that you think will be useful for the radio station?

STAG
9

55

▌▐▐▌ EXERCISE 14.2H

1 You need to obtain a representative sample of 1000 people for an investigation into how often people eat out at restaurants. Comment on the following methods for obtaining the sample.

   a) By choosing 1000 names from the telephone directory.

   b) By stopping 1000 people at random outside the railway station.

   c) By asking 100 restaurants to supply 10 names each.

2 Which sampling method would you use for each of these situations? Describe how you would select the samples.

   a) How long each week is spent on homework by students in your year.

   b) The average number of matches in a box.

   c) The average number of children in the families of students in your school.

▌▐▐▌ EXERCISE 14.3H

1 Rhian wants to conduct an investigation to find out how much time Year 7 students spend watching TV.
At her school there are five classes in Year 7, each of 30 students.

   a) How should she obtain a stratified sample of 10% of Year 7?

   In each class, 40% are boys and 60% are girls.

   b) How will this affect the sample?

2 In a large company there are four departments.
These are the number of employees in each of the departments.

Department	Number of employees
A	175
B	50
C	250
D	125

The directors wish to consult the employees about a proposal to change the number of hours a week they work.
They decide to select a sample of 50 to interview.
How would you select a stratified sample so that workers from all departments are fairly represented?

▌▌▌ EXERCISE 14.4H

1 To monitor the number of birds of a particular species, 100 are trapped and tagged.
The next year a sample of 60 birds of the same species are caught. 24 of them are found to be tagged.
Calculate an estimate of the size of the population of this species of bird.

2 An ornithologist wishes to know the number of bird nests on the face of a cliff. It is not realistic to count every single nest so he decides to estimate the number.

He divides the area into 36 squares and decides to pick 9 squares at random as a sample.

Method

Throw a dice to select a square.

The first throw gives the 'across' square and the second throw the 'up' square.

Count the number of nests in the chosen square.

Repeat the process to get the number of nests in 9 *different* squares.

Multiply the total by 4 to get an estimate of the total number of nests.

Try this two or three times.

How do your answers compare?

Ask others in the class what answers they got.

How do they compare with the exact number of nests?

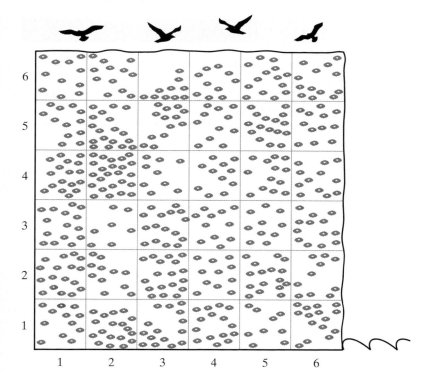

3 The following are the heights, in centimetres, of 100 tomato plants grown by a farmer.

98	125	77	102	105	107	112	108	102	89
115	96	119	86	85	108	89	88	100	100
103	95	99	105	107	103	117	102	92	111
97	94	83	90	104	110	96	115	105	94
122	96	94	103	90	97	101	91	95	100
110	116	96	101	102	113	106	110	126	117
130	102	107	108	112	110	113	94	125	91
105	115	88	118	104	108	116	111	109	112
108	113	106	104	110	128	116	111	122	115
113	122	124	100	120	106	121	103	100	121

To find an estimate of the mean height of the plants, choose a random sample of 20% of the plants (1 in every 5).

Method
Throw a dice to get a number between 1 and 5.
(If you get a 6, throw the dice again.)
Start at this value in the list, counting from the top left-hand corner. Include this start value in your sample.
Count on in 5s along the list.
Write down the height of each value you land on.
Carry on until you have a sample of 20 values.

a) Find the mean height of your sample.

b) Roll the dice again to find a different starting point.
Repeat the method above to find another systematic sample.
Find the mean height of this sample.

c) Choose, at random, another starting point in the grid of numbers.
Find another systematic sample.
Find the mean height of this sample.

d) How do these values of the mean height compare?

e) How does the mean height from your samples compare with the mean height of the whole 100 plants?

EXERCISE 14.5H

1 Comment on the method of sampling in each of these cases.

a) To find out how much support there is for a local football club, an interviewer stops the first ten people leaving a football match.

b) To find out how long cars stay in a pay-and-display car park, a researcher reads the time on the ticket of every tenth car in the car park.

c) A quality control technician takes samples from the production line at the beginning and end of each shift.

2 A researcher decides to take a sample of 200 people to find out what proportion of the population own a car.
Comment on these possible methods for choosing the sample.

a) Ask people returning to a car park in the evening.

b) Select names at random from the electoral register.

c) Ask people at random at a bus station one morning.

Stage 10 Contents

STAGE

10

Using graphs to solve equations

You may find the example useful for Exercises 1.1H and 1.2H.

EXERCISE 1.1H

1 a) Draw the graph of $y = x^2 - 6x + 8$ for $x = 0$ to 6.

   b) On the same grid, draw the line $y = 2x - 5$.

   c) Write down the coordinates of the points where the curve and the line cross.

2 Use the graph of $y = x^2 - 6x + 8$ from question **1**. By drawing another line on the graph, solve the simultaneous equations $y = x^2 - 6x + 8$ and $2y + x = 8$ graphically.

3 Solve these simultaneous equations graphically.

   a) $y = 2x^2 - 1$
      $x + y = 3$

   b) $y = x^2 + x$
      $y = x + 2$

STAG

1C

EXERCISE 1.2H

1 The table of values below is for the equation $y = x^2 - x - 4$.

x	$^-2$	$^-1$	0	1	2	3
y	2	$^-2$	$^-4$	$^-4$	$^-2$	2

a) Draw the graph for values of x from $^-2$ to 3.

b) Use your graph to solve these.
   (i) $x^2 - x - 4 = 0$
   (ii) $x^2 - x - 4 < {}^-2{\cdot}5$

2 a) Draw the graphs of $y = x^2$ and $y = 5 - \frac{1}{2}x$ for $x = {}^-3$ to 3.

b) What is the equation of the points where they intersect?

c) Use your graphs to solve the equation.

3 a) Draw the graph of $y = x^3 - 4x$ for $x = {}^-3$ to 3.

b) Draw another graph so that the equation at the point of intersection is $x^3 - 5x + 3 = 0$.

c) Use your graphs to solve the equation.

4 a) Draw the graph of $y = 2x^2 + 3x - 9$ for values of x from $^-3$ to 2.

b) Use your graph to solve these equations.
   (i) $2x^2 + 3x - 9 = {}^-1$
   (ii) $2x^2 + 3x - 4 = 0$

5 a) Draw the graph of $y = x^2 - 5x + 3$ for values of x from $^-2$ to 8.

b) Use your graph to solve these equations.
   (i) $x^2 - 5x + 3 = 0$
   (ii) $x^2 - 5x + 3 = 5$
   (iii) $x^2 - 7x + 3 = 0$

6 a) Draw the graph of $y = x^2 - 3x$ from $x = {}^-2$ to 5.

b) Use your graph to solve these equations.
   (i) $x^2 - 3x - 2 = 0$
   (ii) $x^2 - 4x + 1 = 0$

Do not draw the graphs in questions **7** and **8**.

7 The intersection of two graphs is the solution to the equation
$x^2 - 2x - 3 = 0$.
One of the graphs is $y = x^2 - x - 1$.
What is the other graph?

8 The graphs of $y = x^2 + 2x$ and $y = 2x + 1$ are drawn on the same grid.
What is the equation whose solution is found at the intersection of the two graphs?

EXERCISE 1.3H

1 a) Draw the graphs of $x^2 + y^2 = 25$ and $y = x + 1$ on the same grid.

b) Find the coordinates of the points where the two graphs cross.

2 Draw graphs to solve these simultaneous equations.

a) $x^2 + y^2 = 4$ and $y = 2x - 1$

b) $x^2 + y^2 = 36$ and $x + y = 5$

3 a) Find the radius of the circle whose equation is $x^2 + y^2 = 17$.

b) Draw graphs to solve the simultaneous equations $x^2 + y^2 = 17$ and $y = x - 3$.

STA
1

Growth and decay

You may find the example helpful for both of these exercises.

1 £3000 is invested at 4% compound interest.

    a) Calculate the value of the investment after
      (i) 2 years.
      (ii) 20 years.

    b) Find a formula for the amount the investment is worth after n years.

2 A car costs £12 000 when new.
  It depreciates in value by 13% per year.

    a) Calculate the value of the car after
      (i) 3 years.
      (ii) 8 years.

    b) Find a formula for the value of the car after n years.

3 The estimated population of a town t years after 2007 is modelled by the formula $P = 95\,000 \times 1 \cdot 1^{-t}$.

  Using this formula,

    a) what was the population in 2007?

    b) what will the population be in 2011?

4 A colony of bacteria is found to increase by 20% every hour.

    a) There are 500 000 bacteria at noon.
      Find the number of bacteria at
      (i) 3 p.m.
      (ii) 6:30 p.m.

    b) Find the formula for the number after n hours.

5 a) Draw the graph of $y = 2 \cdot 5^x$ for values of x from $^-2$ to 4.

   b) Use your graph to estimate
     (i) the value of y when $x = 3 \cdot 4$.
     (ii) the value of x when $y = 11$.

6 a) Copy and complete this table of values for the equation $y = 0 \cdot 8^{-x}$.

x	y
$^-8$	0·17
$^-6$	
$^-4$	
$^-2$	0·64
0	1
2	
4	
6	
8	5·96

   b) Draw the graph of $y = 0 \cdot 8^{-x}$ for values of x from $^-8$ to 8.

   c) Use your graph to solve these equations.
     (i) $0 \cdot 8^{-x} = 0 \cdot 5$
     (ii) $0 \cdot 8^{-x} = 5$

STA

1

EXERCISE 2.2H

1 a) Copy and complete this table for the function $y = 4^x$.

x	y
0	1
1	
2	
3	
4	
5	

b) Find the value of y when $x = 10$.

c) Use trial and improvement to find, correct to 2 decimal places, the value of x when $y = 10\,000$.

2 a) If $y = 2 \cdot 5^x$, calculate the value of y when
   (i) $x = 3$.
   (ii) $x = 4$.

b) Use trial and improvement to solve $2 \cdot 5^x = 20$ accurate to 2 decimal places.

3 A population of butterflies is declining at 8% per year.
The population in August 2001 was 850.

a) Explain why a suitable equation for the population is given by $y = 850 \times 0 \cdot 92^t$.

b) Find the population in August 2006.

c) In August of which year is the population first below 400?

4 Use trial and improvement to solve these equations.
Give your answers to 2 decimal places.

a) $2^x = 14$

b) $5^x = 2$

c) $3^{-x} = 0 \cdot 1$

d) $0 \cdot 5^x = 0 \cdot 1$

e) $2^{-x} = 0 \cdot 1$

5 A car cost £6000 when new.
It depreciates in value by 15% each year.

   a) Explain why this can be represented by the equation
$y = 6000 \times 0.85^x$.

   b) Use trial and improvement to find how old the car is when it is
worth £3000.

6 Mark invests £25 000 at 5·25% p.a. compound interest.
How long will it take for his money to double?

STA

1

3 Rational and irrational numbers

1 State whether each of these numbers is rational or irrational, showing how you decide.

a) 0.1

b) $0.\dot{1}2\dot{3}$

c) $\dfrac{5\pi}{2}$

d) $0.5\dot{4}$

e) $\sqrt{144}$

f) $\sqrt{66}$

g) $5 + 2\sqrt{3}$

h) $\dfrac{5}{6}$

2 State, giving your reasons, whether each of these fractions will terminate or recur.

a) $\dfrac{1}{24}$

b) $\dfrac{7}{25}$

c) $\dfrac{107}{128}$

d) $\dfrac{425}{544}$

e) $\dfrac{32}{75}$

3 Find the fractional equivalent of each of these terminating decimals.
Write each fraction in its simplest form.

a) 0.38

b) 0.504

c) 0.5625

d) $0.531\,25$

4 Convert each of these fractions to recurring decimals using the dot notation.

a) $\dfrac{7}{9}$

b) $\dfrac{5}{18}$

c) $\dfrac{17}{303}$

d) $\dfrac{5}{11}$

e) $\dfrac{5}{7}$

f) $\dfrac{14}{111}$

STAGE
10

5 Convert these recurring decimals to fractions or mixed numbers in their lowest terms.

a) $0 \cdot \dot{7}\dot{2}$

b) $0 \cdot 4\dot{8}$

c) $0 \cdot 3\dot{0}\dot{6}$

d) $0 \cdot 1\dot{2}\dot{3}$

e) $1 \cdot \dot{2}\dot{7}$

f) $1 \cdot 3\dot{8}$

g) $0 \cdot 857 14\dot{2}$

EXERCISE 3.2H

1 Simplify the following, stating whether the result is rational or irrational.

a) $\sqrt{28}$

b) $\sqrt{63}$

c) $\sqrt{125}$

d) $\sqrt{600}$

e) $\sqrt{8} \times \sqrt{50}$

f) $\sqrt{75} \div \sqrt{27}$

g) $\sqrt{32} \times \sqrt{8}$

h) $\sqrt{15} \times \sqrt{27}$

i) $\dfrac{\sqrt{15}}{\sqrt{27}}$

j) $\dfrac{\sqrt{10} \times \sqrt{12}}{\sqrt{15}}$

k) $\sqrt{18} + \sqrt{72}$

2 If $x = 5 + \sqrt{2}$ and $y = 5 - \sqrt{2}$, simplify these.

a) $x + y$

b) $x - y$

c) xy

3 If $x = 7 + \sqrt{3}$ and $y = 5 - 2\sqrt{3}$, simplify these.

a) $x + y$

b) $x - y$

c) x^2

d) xy

4 Rationalise the denominator in each of these irrational fractions. Simplify your answer where possible.

a) $\dfrac{1}{\sqrt{3}}$

b) $\dfrac{3}{\sqrt{5}}$

c) $\dfrac{7}{\sqrt{10}}$

d) $\dfrac{10}{\sqrt{5}}$

e) $\dfrac{5}{2\sqrt{3}}$

f) $\dfrac{1}{3\sqrt{2}}$

g) $\dfrac{\sqrt{3}}{3\sqrt{2}}$

h) $\dfrac{7\sqrt{5}}{5\sqrt{7}}$

5 Find the exact value of x, expressing your answer as simply as possible.

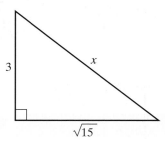

6 In this diagram angle ACB is 30°.

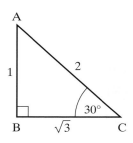

a) Use the diagram to show that
  (i) sin 30° is a rational number.
  (ii) cos 30° is an irrational number.

b) What can you say about tan 30°?

Trigonometry in non-right-angled triangles

1 Find c, B and b.

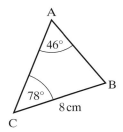

2 Find C, B and b.

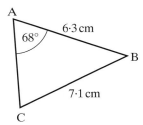

3 Find C, A and a.

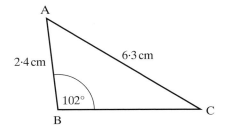

STAG

1(

4 Find A, b and c.

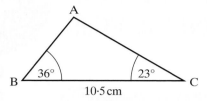

5 Find A, C and c.

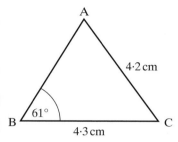

6 In triangle EFG, EF = 7 cm, FG = 9 cm and angle G = 39°.
Find these.

    a) Angle E **b)** Angle F **c)** Side EG

7 In triangle PQR, PQ = 7·8 cm, angle R = 79° and angle P = 51°.
Find these.

    a) Side QR   **b)** Angle Q **c)** Side PR

8 ABC is a flower bed in a garden.
AB is 3·56 m long.

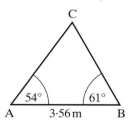

Find the length of the other two sides.

9 The angles of elevation of the bottom (B) and top (T) of a flagpole on top of a tower (AB) are 37° and 42°.
These are measured 200 m from the base of the tower at a point C.

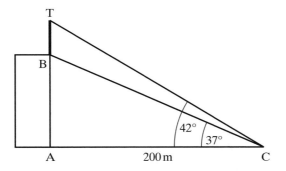

a) Find the length BC.

b) Find the height of the flagpole, BT.

10 P, Q and R are three buoys marking a sailing course.
The bearing of P from Q is 035°
The bearing of P from R is 310°
The bearing of R from Q is 075°
The length of QR is 5·3 km.

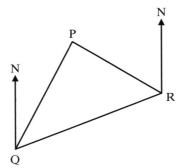

Find the total length of the three stages of the course from P back to P.

EXERCISE 4.2H

1 Find length AC.

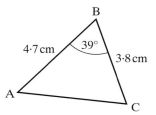

2 Find length AB.

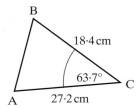

3 Find angle CAB.

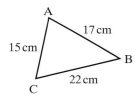

4 Find angle CAB.

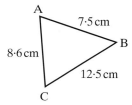

5 In triangle PQR, PR = 7·2 cm, QR = 6·3 cm and angle PRQ = 37°.
Find these.

   a) The length of PQ   **b)** Angle PQR

6 The lengths of the sides of triangle ABC are AB = 4·6 cm,
BC = 11·5 cm and CA = 7·8 cm.
Find the sizes of the angles of the triangle.

7 Find these lengths.

  a) BD

  b) AB

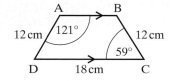

8 Find these lengths.

  a) BC

  b) AD

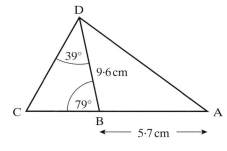

9 ABCDEFGH is a cuboid with lengths of 2 cm, 4 cm and 1 cm.

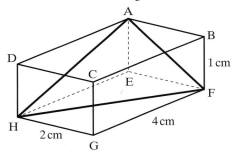

Find the angles of triangle AFH.

10 A, B and C are points on an orienteering course.
The bearing of B from A is 040°.
The bearing of C from B is 125°.
AB is 3·7 km and BC is 2·3 km.

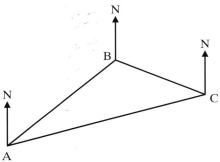

How long is AC?

STA

1

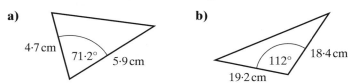

1 Find the area of each of these triangles.

a)

4·7 cm 71·2° 5·9 cm

b)

19·2 cm 112° 18·4 cm

2 In triangle ABC, AB = 15·3 cm, AC = 9·6 cm and angle BAC = 53·6°.
Find the area of triangle ABC.

3 ABCD is a kite.

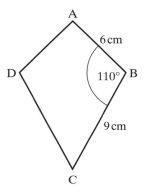

Find the area of the kite.

4 ABCD is a children's playground.
OA = 83 m, OB = 122 m, OC = 106 m and OD = 78 m.

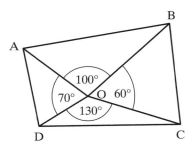

Calculate the area and the perimeter of the playground.
Give your answers to 3 significant figures.

AGE

0

5 O is the centre of the circle. Angle AOB = 120°.

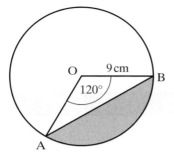

Find the area of the shaded segment.

5 Trends and time series

EXERCISE 5.1H

1 The table shows the number of copies done on a
school photocopier for the first three weeks of term.

	M	Tu	W	Th	F
Wk 1	720	560	430	280	220
Wk 2	680	490	420	260	200
Wk 3	560	430	380	190	180

a) Plot these figures on a graph.
Use a scale of 1 cm to each day on the horizontal
axis and 2 cm to 100 copies on the vertical axis.

b) Calculate the 5-day moving averages.

c) Plot the moving averages on your graph.

d) Comment on the daily variation and the general
trend and suggest an explanation for these.

2 The table opposite shows the number of houses sold
by an estate agent over a 3-year period.
The table also shows some of the 3-monthly moving
averages for the sales.

Year	Month	Sales	3-month moving average
2003	September	15	
	October	14	14·0
	November	13	13·7
	December	14	13·3
2004	January	13	13·7
	February	14	14·7
	March	17	15·3
	April	15	16·7
	May	18	17·3
	June	19	17·7
	July	16	16·7
	August	15	15·7
	September	16	14·7
	October	13	14·3
	November	14	13·7
	December	14	14·0
2005	January	14	15·0
	February	17	16·7
	March	19	17·3
	April	16	17·7
	May	18	17·7
	June	19	18·3
	July	18	18·0
	August	17	17·3
	September	17	16·3
	October	15	16·0
	November	16	15·3
	December	15	16·0
2006	January	17	17·0
	February	19	18·7
	March	20	20·3
	April	22	
	May	20	
	June	19	
	July	19	
	August	14	

a) Work out the last four moving averages.

b) Draw a graph to show the sales and the moving averages.

c) Comment on the quarterly variation.

STA

1

3 The table shows the number of people visiting a dentist's surgery each day of a 4-week period.

	M	Tu	W	Th	F
Wk 1	35	42	63	24	51
Wk 2	33	24	60	26	48
Wk 3	38	44	66	29	46
Wk 4	35	47	65	27	49

a) Plot a time-series graph of these figures.

b) One of the dentists works part-time.
On which day does he not work?

c) On one day in this period, a dentist was ill and her appointments were postponed.
When was this?

d) Calculate the 5-day moving averages and add these to the graph.

4 The graph on the next page shows the quarterly sales figures, in millions of pounds for 2003 to 2006 for a company, together with the 4-point moving averages.

a) Use the graph to calculate the last two moving averages, which have not been plotted.

b) Comment on the quarterly variation and the general trend.

c) Explain why it might not be reliable to use the graph to predict what will happen next year.

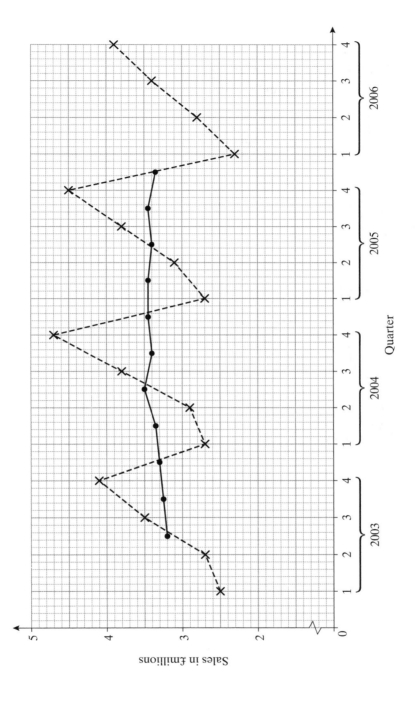

5 The graph is a speed–time graph for the journey of a car.

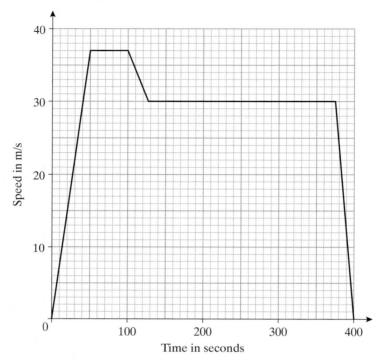

a) Describe the car's journey.

b) Calculate the acceleration in the first 50 seconds.

c) Calculate the retardation in the last 25 seconds.

6 Peter decided to take a bath.
He turned on the hot tap but after a while he decided that the
water was too hot so he turned on the cold tap as well.

When the bath was full enough, he turned both taps off and got in.
After 5 minutes he decided that the water was too cool so he
pulled the plug out for half a minute and then topped up with hot
water.

After a further 10 minutes Peter got out and pulled the plug out to
empty the bath.
Each tap flows at 5 gallons of water per minute and when Peter
has the bath it contains 25 gallons.
The outflow lets water out at a rate faster than that of one tap but
not as fast as two.

Draw a possible graph to show the number of gallons of water in
the bath at any given time.
Use a scale of 2 cm to 5 gallons on the vertical axis and 2 cm to 5
minutes on the horizontal axis.

7 Anne, Bethany and Catherine run a 10 km race.
Their progress is shown in the graph.

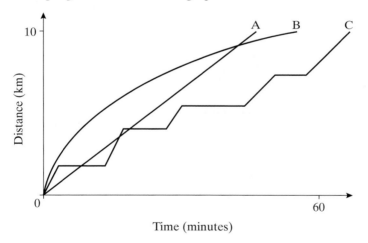

Imagine you are a commentator and give a description of the race.

Congruency – proving and using

EXERCISE 6.1H

pt

1 Which of these pairs of triangles are congruent?
Explain your answer in each case.

a)

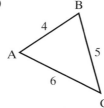

b)

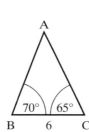

c)

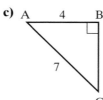

d)

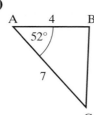
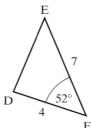

2 ABCD is a kite.
AB = AD and CD = CB.

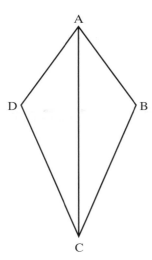

a) Prove that triangle ABC is congruent to triangle ADC.

b) What conclusion can you make about the opposite angles of a kite?

3 Use congruent triangles to prove that the diagonals of a parallelogram bisect each other.

4 ABCD and APQR are squares.

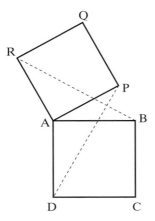

Prove that triangles RAB and PAD are congruent.

STA

1(

5 A, B and C are points on a circle.
AB = BC and PB bisects angle ABC.
Prove that PA = PC.
Does this show that P is the centre of the circle?

6 ABC is a triangle with AB = AC.
D is the midpoint of BC.
Prove that AD bisects angle BAC.

7 The diagram consists of two parallel lines and two other straight lines.
BE = EC.
Prove that AB = CD.

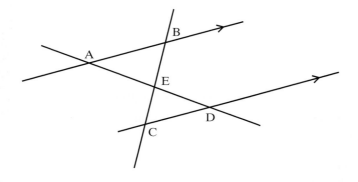

Calculating the roots of equations

Solve these quadratic equations by completing the square.
Give your answers correct to 2 decimal places.

1 $x^2 + 4x - 10 = 0$

2 $x^2 - 6x + 1 = 0$

3 $x^2 + 12x - 5 = 0$

4 $x^2 - 10x - 2 = 0$

5 $x^2 - 3x - 7 = 0$

6 $x^2 + 18x - 35 = 0$

7 $x^2 - 7x + 2 = 0$

8 $x^2 - 9x - 35 = 0$

9 $x^2 + 3x - 9 = 0$

10 $x^2 + 4x - 17 = 0$

STAG

1C

EXERCISE 7.2H

Use the formula to solve these equations.
Give your answers correct to 2 decimal places.

1 $x^2 + 7x + 3 = 0$

2 $2x^2 - x - 4 = 0$

3 $5x^2 + 11x + 1 = 0$

4 $3x^2 - x - 8 = 0$

5 $x^2 - 6x + 1 = 0$

6 $x^2 + 12x - 5 = 0$

7 $9x^2 + 6x - 19 = 0$

8 $x^2 + 3x + 1 = 0$

9 $x^2 + x - 1 = 0$

10 $x^2 - 2x - 4 = 0$

11 $2x^2 - 4x + 1 = 0$

12 $3x^2 + 8x - 9 = 0$

Surface areas and complex shapes

8

EXERCISE 8.1H

1 Calculate the curved surface area of a cylinder with these dimensions.

   a) Radius = 6·1 cm, height = 8·9 cm

   b) Radius = 0·4 mm, height = 18 mm

2 Calculate the curved surface areas of each of these cones.

   a)

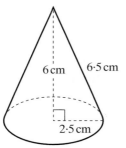

6 cm   6·5 cm

2·5 cm

   b)

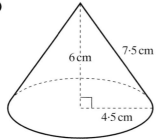

6 cm   7·5 cm

4·5 cm

3 Calculate the surface area of a sphere of each of these radii.

   a) 4·6 cm

   b) 8 mm

STAG

10

4 Find the total surface area of
this solid cylinder.

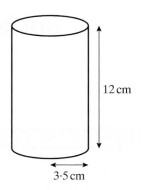

12 cm

3·5 cm

5 A cone has a radius of 7 cm and a perpendicular height of 15 cm.

   a) Find the slant height of the cone.
      Give your answer to 2 decimal places.

   b) Find the curved surface area of the cone.
      Give your answer to the nearest whole number.

6 A sphere has a total surface area of 200 cm^2.
Find its radius.

7 The diagram shows a pyramid with a square base of side 9 cm.
The perpendicular height of the pyramid is 10 cm.

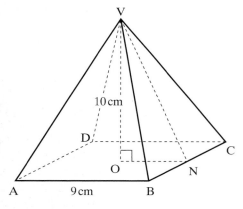

   a) Calculate the length of VN.

   b) Calculate the total surface area of the pyramid.

AGE

0

8 A cone and a cylinder have equal bases and equal heights.

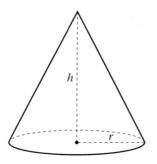 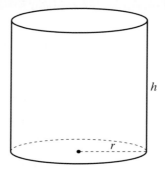

a) If the volume of the cylinder is 36 cm³, find the volume of the cone.

b) If the volume of the cone is 19 cm³, find the volume of the cylinder.

c) If the combined volume of both shapes is 64 cm³, find the volume of the cylinder.

9 The diagram shows a solid cylindrical metal rod with a diameter of 2 cm and a height of 4 cm standing inside a cylinder with a diameter of 3 cm.
The cylinder contains oil to a depth of 3 cm.

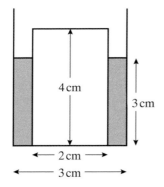

a) Find the volume of the oil poured into the cylinder.

b) Find the volume of the extra oil that would have to be added to just cover the rod.

c) Oil is actually added to a depth 2 cm greater than the height of the rod.
What is the total volume of oil in the cylinder?

EXERCISE 8.2H

1 A solid cylinder has a base radius of 5 cm and a volume of 350 cm³.
Calculate its curved surface area.

2 Calculate

   a) the length of the chord AB.

   b) the perimeter of the shaded segment.

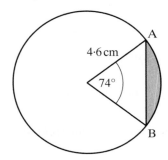

3 Calculate the area of the shaded segment.

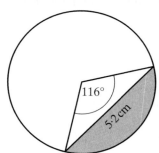

4 A cone of height 12 cm and base radius 8 cm has a cone of height 3 cm removed from its top as shown.

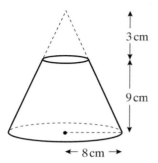

a) Find the radius of the base of the top cone.

b) Calculate the volume of the frustum.

5 A cylindrical glass bowl of radius 15 cm has water in it with floating candles. 20 glass marbles of radius 1·2 cm are placed in the bowl.

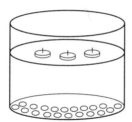

By how much does the water level in the bowl increase?

6 A cone has a base radius of 4·3 cm and height of 8·4 cm.
It has the same volume as a sphere.
Find the radius of the sphere.

7 A hemispherical glass bowl has an internal diameter of 22 cm and is 7·5 mm thick throughout.
5% of the original glass is removed when the pattern is cut into the bowl.
Calculate the volume of glass remaining.

STA

1

8 A salt pot is in the form of a hollow cylinder of diameter 3 cm and height 3 cm with a hemispherical shell fixed on top.
Salt is poured into the pot to a depth of 2·5 cm.
The pot is inverted, with the hole covered, so the flat base of the cylinder is horizontal.

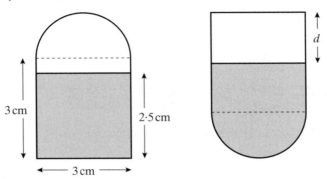

Find the distance, d, from the top of the salt to the flat base of the cylinder.

Working with algebraic fractions

9

You may find the example helpful for both of these exercises.

EXERCISE 9.1H

Simplify these.

1 $\dfrac{a}{2} + \dfrac{a+1}{3}$

2 $\dfrac{2x-1}{5} + \dfrac{x+3}{2}$

3 $\dfrac{x-3}{3} - \dfrac{x-2}{5}$

4 $\dfrac{1}{b} + \dfrac{2}{b+1}$

5 $\dfrac{5}{x-2} + \dfrac{3}{x+3}$

6 $\dfrac{p-1}{3} + \dfrac{p+2}{4}$

7 $\dfrac{p+1}{3} - \dfrac{2p+1}{4}$

8 $\dfrac{2x+1}{7} - \dfrac{x+2}{3}$

9 $\dfrac{3}{x-2} + \dfrac{4}{x}$

10 $\dfrac{7}{m+1} - \dfrac{3}{m+2}$

11 $\dfrac{x+1}{x+2} + \dfrac{x+2}{x+1}$

12 $\dfrac{1}{x} - \dfrac{1}{x-1} + \dfrac{1}{x+1}$

▌▐▐▐ EXERCISE 9.2H

Solve these.

1 $\dfrac{x}{3} + \dfrac{2x}{5} = 1$

2 $\dfrac{x}{7} + \dfrac{1}{3} - \dfrac{2}{7} = \dfrac{^-2}{3}$

3 $\dfrac{x+1}{4} + \dfrac{x-1}{3} = \dfrac{17}{12}$

4 $\dfrac{2(x-1)}{5} - \dfrac{3(1-x)}{2} = \dfrac{19}{10}$

5 $\dfrac{x}{3} - \dfrac{x-1}{4} = 7$

6 $\dfrac{1}{x} + \dfrac{x}{6} = \dfrac{7}{6}$

7 $\dfrac{4}{x} + \dfrac{3}{2x} = \dfrac{11}{4}$

8 $x + 2 = \dfrac{x+2}{x-3}$

9 $\dfrac{x+1}{4} - \dfrac{x-1}{5} = 1$

10 $\dfrac{2}{3x} - \dfrac{6}{5x} = {}^-8$

11 $\dfrac{1}{x+1} + \dfrac{1}{x-1} = 4$

12 $(x-2)(x+5) = 6x + 8$

Vectors

EXERCISE 10.1H

1 Write down the column vectors for $\overrightarrow{AB}$, $\overrightarrow{AD}$, $\overrightarrow{CB}$, $\overrightarrow{DC}$ and $\overrightarrow{AC}$.

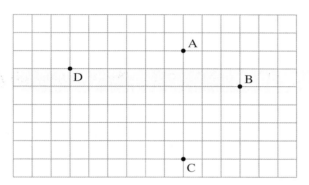

2 Find the column vector that maps

   a) $(1, 3)$ on to $(2, 5)$.

   b) $(^-2, 0)$ on to $(3, ^-1)$.

   c) $(^-5, 2)$ on to $(^-5, ^-5)$.

   d) $(3, 4)$ on to $(0, 0)$.

3 Find where the point is mapped by the vector.

   a) $(2, 3)$ by $\begin{pmatrix} 5 \\ 1 \end{pmatrix}$

   b) $(3, ^-2)$ by $\begin{pmatrix} 0 \\ 5 \end{pmatrix}$

   c) $(^-2, 4)$ by $\begin{pmatrix} ^-1 \\ 3 \end{pmatrix}$

   d) $(^-5, ^-2)$ by $\begin{pmatrix} 3 \\ ^-4 \end{pmatrix}$

STAG

1C

4 Write down the vectors $\overrightarrow{PQ}$, $\overrightarrow{RS}$, $\overrightarrow{TU}$, $\overrightarrow{VW}$ and $\overrightarrow{XY}$ in terms of **a** or **b**.

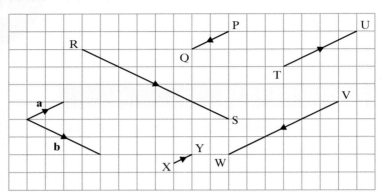

▌▌▌ EXERCISE 10.2H

1 Work out these.

a) $3 \times \begin{pmatrix} 1 \\ 2 \end{pmatrix}$
b) $\begin{pmatrix} 3 \\ 4 \end{pmatrix} + \begin{pmatrix} 1 \\ 6 \end{pmatrix}$
c) $\begin{pmatrix} 3 \\ 4 \end{pmatrix} - \begin{pmatrix} 1 \\ 6 \end{pmatrix}$

2 Given that $\mathbf{a} = \begin{pmatrix} -3 \\ 0 \end{pmatrix}$, work out these.

a) $3\mathbf{a}$
b) $-\mathbf{a}$
c) $-\frac{1}{3}\mathbf{a}$

3 Given that $\mathbf{a} = \begin{pmatrix} 1 \\ 2 \end{pmatrix}$, $\mathbf{b} = \begin{pmatrix} 1 \\ 3 \end{pmatrix}$, work out these.

a) $\mathbf{a} - \mathbf{b}$

b) $2\mathbf{a} + 3\mathbf{b}$

c) $3\mathbf{a} - 2\mathbf{b}$

4 Given that $\mathbf{a} = \begin{pmatrix} 2 \\ 0 \end{pmatrix}$, $\mathbf{b} = \begin{pmatrix} -3 \\ 0 \end{pmatrix}$ and $\mathbf{c} = \begin{pmatrix} 0 \\ -4 \end{pmatrix}$, work out these.

a) $\mathbf{a} - \mathbf{b} - \mathbf{c}$

b) $2\mathbf{a} - \mathbf{b} + 3\mathbf{c}$

c) $\frac{1}{2}\mathbf{a} + \frac{1}{2}\mathbf{b} + \frac{1}{2}\mathbf{c}$

EXERCISE 10.3H

pt

1 The diagram shows the vectors **a** and **b**.

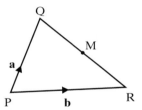

Copy the diagram and draw the resultant of 2**a** + 2**b**.

2 In the triangle, $\overrightarrow{PQ}$ = **a** and $\overrightarrow{PR}$ = **b**.

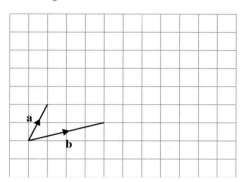

a) Work out the vector $\overrightarrow{QR}$ in terms of **a** and **b**.

b) M is the midpoint of QR.

   Find the vector $\overrightarrow{PM}$ in terms of **a** and **b**.

3 PQRS is a parallelogram.
The midpoints of PQ, QR, RS and SP are I, J, K and L respectively.
$\overrightarrow{SR} = \mathbf{a}$ and $\overrightarrow{RQ} = \mathbf{b}$.

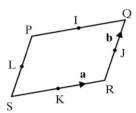

a) Write these vectors in terms of **a** and/or **b**.

(i) $\overrightarrow{PQ}$ (ii) $\overrightarrow{PS}$

(iii) $\overrightarrow{SQ}$ (iv) $\overrightarrow{LI}$

b) What do the vectors show about LI and SQ?

4 ABCD is a rectangle.
The midpoints of AB, BC, CD and DA are E, F, G and H respectively.
$\overrightarrow{AB} = \mathbf{p}$ and $\overrightarrow{BC} = \mathbf{q}$.

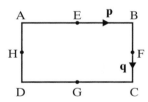

a) Write these vectors in terms of **p** and/or **q**.

(i) $\overrightarrow{EF}$ (ii) $\overrightarrow{HG}$

(iii) $\overrightarrow{EH}$ (iv) $\overrightarrow{FG}$

b) What does this tell you about EFGH?

5 Triangle OAB is enlarged, with centre O and scale factor ⁻2 to give triangle OCD.

$\overrightarrow{OA} = \mathbf{a}$ and $\overrightarrow{OB} = \mathbf{b}$.

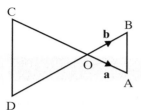

Use vectors to show that CD is parallel to BA.

6 OACB is a parallelogram.

$\overrightarrow{OA} = \mathbf{a}$ and $\overrightarrow{OB} = \mathbf{b}$.

D is the midpoint of OC and E is the midpoint of AB.

a) Write these vectors in terms of **a** and/or **b**.

    (i) $\overrightarrow{OD}$ **(ii)** $\overrightarrow{OE}$

b) What can you deduce from your answer to part **a)**?

7 AP = 2OA and BQ = 2OB.

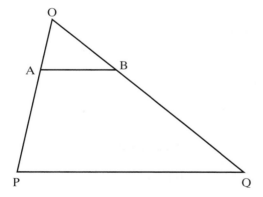

a) Use vectors to prove that PQ = 3AB.

b) What else can you deduce?

Comparing sets of data

EXERCISE 11.1H

1 The lengths of 200 leaves from a variety of trees were measured at each of two sites, one exposed and one sheltered.
The results are shown in the table.

Length of leaf (l cm)	Frequency	
	Exposed site	Sheltered site
$7 < l \leqslant 8$	8	0
$8 < l \leqslant 9$	25	2
$9 < l \leqslant 10$	62	14
$10 < l \leqslant 11$	53	31
$11 < l \leqslant 12$	32	88
$12 < l \leqslant 13$	17	42
$13 < l \leqslant 14$	2	18
$14 < l \leqslant 15$	1	5

Draw a cumulative frequency diagram showing both these distributions.
Compare the distributions.

2 The distribution of times taken to run a marathon race are shown in the table.

Time (t minutes)	Frequency
$120 < t \leqslant 130$	2
$130 < t \leqslant 140$	10
$140 < t \leqslant 150$	28
$150 < t \leqslant 160$	39
$160 < t \leqslant 170$	75
$170 < t \leqslant 180$	47
$180 < t \leqslant 190$	26
$190 < t \leqslant 200$	13

a) Draw a cumulative frequency diagram, with a box plot below it, to represent this distribution.

Another marathon race had these summary data.

Number in race	80
Lower quartile	150 minutes
Median	158 minutes
Upper quartile	164 minutes

b) Make three comparisons between the two races.

STA

1

3 This histogram shows the ages of people consulting the doctors at a health centre on a day in February.

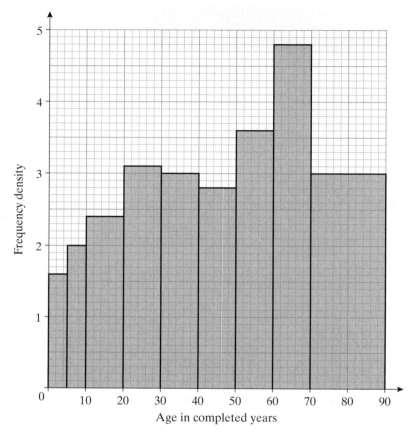

a) The left-hand bar actually represents people aged 0–4 years old. Explain why it extends from 0 to 5 years.

b) How many people aged 0–4 years old visited the doctors?

The table shows the ages of people visiting the dentist on the same day.

Age in completed years	Frequency
0–4	12
5–9	35
10–19	42
20–29	36
30–39	28
40–49	18
50–59	12
60–89	6

c) Draw a histogram for this data.

d) Make two comparisons about the ages of people visiting the doctor and the dentist.

4 These box plots show the ages of people taking part in competitive team sports in the years 1990 and 2000.

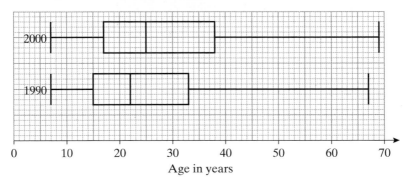

Make three comparisons between the two years.
(Your comparisons should include numerical data.)

5 These two histograms show the ages of the passengers on two buses one morning.

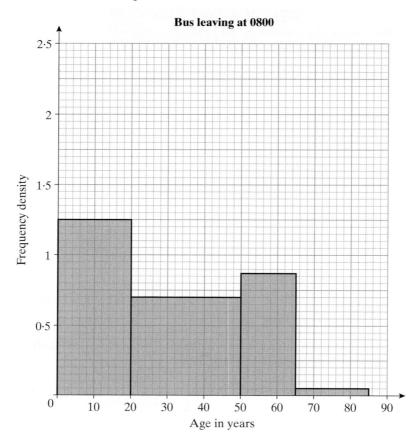

Bus leaving at 0800

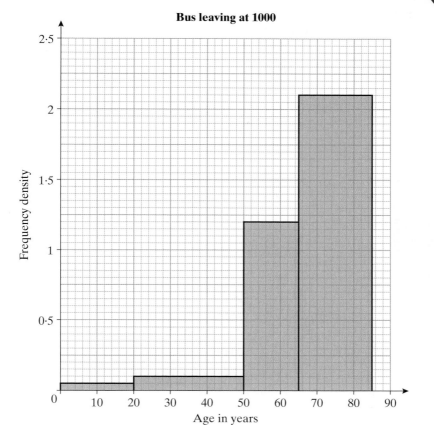

Bus leaving at 1000

a) Comment on the two distributions.

b) Calculate the mean age of the passengers on each bus.
Give your answers to 1 decimal place.

6 The cumulative frequency diagram shows the lifetimes of two types of lightbulb, A and B.

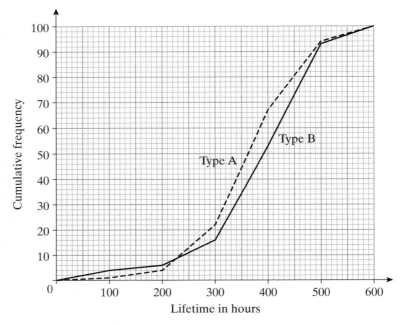

Compare the performances of the two types of lightbulb using the median and the interquartile range.

Simultaneous equations

Solve these simultaneous equations by the method of substitution.

1 $2x + 3y = 5$
   $x - y = 10$

2 $x + y = 2$
   $3x + 2y = 2$

3 $2x - y = 5$
   $3x + 2y = \frac{1}{2}$

4 $5x - y = {}^-5$
   $2x + 3y = {}^-19$

5 $x + y = 3$
   $3x + 5y = 16$

Solve these simultaneous equations by the method of substitution.

1 $y + x - 3 = 0$
   $y = x^2 + 1$

2 $y = x^2 + x$
   $y = x + 1$

3 $y = x^2 + 3x - 1$
   $x - 2y - 4 = 0$

4 $y = 7x - 10$
   $y = x^2$

5 $x + y = 3$
$4x - y^2 = 0$

6 $y = 4x + 3$
$y = x^2 - 3x - 5$

7 $y - x = 5$
$y = x^2 - 2x + 1$

8 $y = 2x^2 + 7x - 6$
$y - 2x = 6$

▌▌▌ EXERCISE 12.3H

Use algebra to solve these simultaneous equations.

1 $x^2 + y^2 = 17$
$y = x - 3$

2 $x^2 + y^2 = 17$
$x + y = 3$

3 $x^2 + y^2 = 5$
$x + y = 3$

4 $x^2 + y^2 = 10$
$x - y = 4$

5 $x^2 + y^2 = 10$
$x + y = 4$

6 $x^2 + y^2 = 2$
$y = 2x - 1$

7 $x^2 + y^2 = 36$
$x + y = 5$

8 $x^2 + y^2 = 9$
$2y - x = 1$

Trigonometrical functions

EXERCISE 13.1H

1 a) Sketch the graph of $y = \sin x$ for values of x from $0°$ to $360°$.

b) Use your graph and your calculator to find the solutions of $\sin x = 0.7071$ between $0°$ and $360°$.

2 a) Sketch the graph of $y = \cos x$ for values of x from $^-360°$ to $360°$.

b) Use your graph and your calculator to find the solutions of $\cos x = 0.866$ between $^-360°$ and $360°$.

3 a) Sketch the graph of $y = \tan x$ for values of x from $^-180°$ to $360°$.

b) Use your graph and your calculator to find the solutions of $\tan x = 0.3640$ between $^-180°$ and $360°$.

4 Given that $\cos 40° = 0.766$, use the symmetry of the graph of $y = \cos x$ to find the solutions of $\cos x = ^-0.766$ between $0°$ and $360°$.

5 Give three other angles that have a sine value equal to each of these.

a) $\sin 30°$ **b)** $\sin {}^-50°$

c) $\sin 45°$ **d)** $\sin {}^-240°$

6 Give three other angles that have a cosine value equal to each of these.

a) $\cos 120°$ **b)** $\cos {}^-90°$ **c)** $\cos {}^-30°$

7 Give three other angles that have a tangent value equal to each of these.

a) $\tan {}^-45°$ **b)** $\tan 60°$ **c)** $\tan 200°$

8 For $0° \leqslant x \leqslant 360°$, solve these equations.

a) $\sin x = 0.1$ **b)** $2 \cos x = 1$

c) $\tan x = {}^-3$ **d)** $\cos x = {}^-\sin x$

EXERCISE 13.2H

1 a) Sketch the graph of $y = 4 \sin x$ for values of x from $0°$ to $360°$.

  b) State the period and amplitude of the graph.

2 a) Sketch the graph of $y = \cos 4x$ for values of x from $0°$ to $180°$.

  b) State the period and amplitude of the graph.

3 a) Sketch the graph of $y = \cos \frac{1}{2}x$ for values of x from $0°$ to $360°$.

  b) State the period and amplitude of the graph.

4 Find the solutions of $5 \sin x = 1$ for x between $0°$ and $360°$.

5 Find the solutions of $\cos 4x = 1$ for x between $0°$ and $360°$.

6 Find the solutions of $\cos \frac{1}{2}x = {}^-1$ for x between $0°$ and $360°$.

7 a) Sketch the graph of $y = \sin 3x$ for x between $0°$ and $360°$.

  b) Find the solutions of $\sin 3x = \frac{1}{2}$ for x between $0°$ and $360°$.

8 Find the solutions of $\tan 2x = 1$ for x between $0°$ and $360°$.

Transforming functions

EXERCISE 14.1H pt

1 a) Sketch these graphs on the same diagram.
  (i) $y = x^2$
  (ii) $y = x^2 - 4$

  b) State the transformation that maps **(i)** on to **(ii)**.

2 a) Sketch these graphs on the same diagram.
  (i) $y = x^2$
  (ii) $y = (x - 3)^2$

  b) State the transformation that maps **(i)** on to **(ii)**.

3 State the equation of the graph after $y = x^2$ has been translated by $\begin{pmatrix} -2 \\ 0 \end{pmatrix}$.

4 a) On the same diagram, sketch the graphs of $y = x^3$, $y = (x - 5)^3$ and $y = (x - 5)^3 + 6$.

  b) State the transformation which maps $y = x^3$ on to $y = (x - 5)^3 + 6$.

5 a) Sketch these graphs on the same diagram.
  (i) $y = \cos x$
  (ii) $y = \cos x - 1$

  b) State the transformation that maps **(i)** on to **(ii)**.

STAG

10

6 This graph is a transformed sine curve.
State its equation.

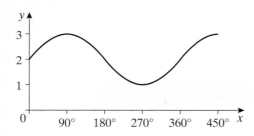

7 This is the graph of $y = f(x)$.

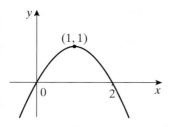

Find the coordinates of the highest points on these graphs.

a) $y = f(x) + 2$

b) $y = f(x - 2)$

8 By completing the square, find the coordinates of the lowest point on the graph of $y = x^2 + 4x - 3$.

▮▮▮ EXERCISE 14.2H

1 a) Sketch on the same diagram the graphs of $y = \sin x$ and $y = 2 \sin x$ for $0° \leqslant x \leqslant 360°$.

b) Describe the transformation that maps $y = \sin x$ on to $y = 2 \sin x$.

2 Describe the transformation that maps $y = \cos x$ on to $y = \cos \frac{1}{2}x$.

3 a) Sketch on the same diagram the graphs of $y = \cos x$ and $y = \cos 4x$ for $0° \leqslant x \leqslant 180°$.

b) Describe the transformation maps $y = \cos x$ on to $y = \cos 4x$.

4 The graph of $y = x^3 + 5$ is reflected in the x-axis.
State the equation of the resulting graph.

AGE
0

5 The graph of $y = x^3 + 2$ is reflected in the y-axis.
State the equation of the resulting graph.

6 State the equation of the graph of $y = x^3 + 2$ after it is stretched

   a) parallel to the y-axis with scale factor 3.

   b) parallel to the x-axis with scale factor $\frac{1}{4}$.

7 On the same axes, sketch the graphs of

   a) $y = 2 \sin x$

   b) $y = \sin 3x$

   c) $y = 2 \sin 3x$

8 This is the graph of $y = a \cos bx$.
Find the values of a and b.

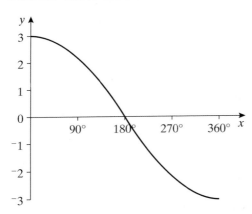

15 Probability

EXERCISE 15.1H

1 It is estimated that 3 out every 10 cars over 15 years old will fail the M.O.T. test because of a problem with their lights.

Of those that pass, it is estimated that 4 out of 10 will then fail on brakes.

Of those that pass the test so far, 25% will then fail because of a problem with steering.

Find the probability that a car over 15 years old will pass these three checks of the M.O.T. test.

2 A box contains seven red pens and three blue pens.
Jane takes a pen from the box and keeps it.
Susan then takes a pen from the box.
Both girls take their pens without looking.

Find the probability that Jane and Susan

a) both take red pens.

b) both take blue pens.

c) take pens of the same colour.

d) take pens of different colours.

3 Two fair six-sided dice are rolled.

What is the probability of getting

a) two prime numbers?

b) a prime number on one dice and a 6 on the other?

4 A bag contains two red counters, three white counters and four blue counters.
Three counters are drawn from the bag without replacement.

Find the probability that

a) they are all red.

b) they are all blue.

c) there is one of each colour.

d) there are at least two of the same colour.

5 The letters of the word PREPOSSESSING are placed in a box. A letter is selected and then replaced in the box and a second letter is then selected.

Find the probability that

a) the letter S is chosen twice.

b) the letter G is chosen twice.

c) a P and an E are chosen, in either order.

6 The probability that it will rain on any day is 0·4.
If it rains one day, the probability that it will rain the next day is 0·7.
If it does not rain that day, the probability that it will rain the next day is 0·2.

Find the probability that

a) it rains for three consecutive days.

b) on three consecutive days it rains at least once.

STA

1